FLIGHT

Malcolm Dixon

Illustrations by Stephen Wheele

Exploring Technology

Bridges and Tunnels
Communications
Flight
Machines
Moving on Land
Moving on Water
Structures
Textiles

For Michelle and Joanne

Series editor: Sue Hadden
Book editor: Joanne Jessop
Designer: Malcolm Walker, Kudos Designs

First published in 1990 by
Wayland (Publishers) Ltd
61 Western Road, Hove
East Sussex BN3 1JD, England

**British Library Cataloguing in
Publication Data**
Dixon, Malcolm, *1946–*
  Flight.
  1. Flight
  I. Title II. Series
  629.132

  ISBN 1–85210–931–9

Phototypeset by Nicola Taylor, Wayland.
Printed in Italy by G. Canale & C.S.p.a.
Bound in France by A.G.M.

Cover: *The invention of the hang glider has given humans the opportunity to glide like a bird.*

CONTENTS

Into the air

Throughout history people have envied birds their ability to fly through the air. From ancient times there have been myths about people who could fly like birds. According to Greek legend, Daedalus and his son Icarus made wings from wax and feathers to help them escape from prison. Daedalus was successful in his escape. But Icarus flew so close to the sun that the wax melted and the feathers dropped off his wings. Icarus fell to the sea and drowned.

Although people have never been able to fly using bird-like wings, over the centuries scientists and engineers have designed flying machines. In the fifteenth century the great Italian artist Leonardo da Vinci made many accurate drawings of birds in flight. From these studies he made designs for flying machines. None of Leonardo's designs were tested, but they did predict machines to come. In the last two hundred years people have been able to design and make many types of flying machines.

The inventors of these machines have been faced with many problems to overcome. On many occasions their designs would not work, and they had to work out why this was so before trying again. The inventors

Leonardo da Vinci drew these sketches for an aircraft in the 15th century.

tried to improve upon previous machines and to learn from the ideas of others. Gradually, they worked out how an aircraft could be made to stay in the air and what were the best materials to use.

This book explains how flight has developed from hot-air balloons, kites and gliders to the modern jet

The British Harrier jump jet can adjust its jet engine for vertical take-off and landing.

aeroplanes and rocket-propelled spacecraft. The experiments and practical instructions are designed to help you understand the basic principles of flight.

Moving air

You need:
Two ping-pong balls
Paper
Cotton thread
Scissors
Sticky tape
Hair dryer
Pencil

First try these experiments:

1. Using cotton and sticky tape, hang two ping-pong balls from a pencil as shown. The balls should be 2 or 3 cm apart. Use a hair dryer to blow air through the gap between them. What happens. Are you surprised?

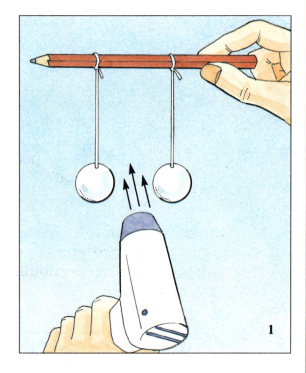

2. Cut a strip of paper and fold it into a bridge shape. Blow gently through the bridge. What happens?

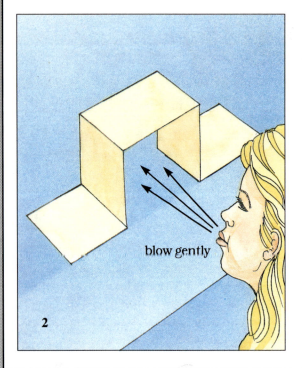

blow gently

Both experiments can be explained by a discovery made in 1738 by a Swiss scientist called Daniel Bernoulli. He found that moving air has less pressure than the still air around it. This is known as the Bernoulli principle. This discovery was very important to the designers of early aircraft.

Air and lift

You need:
Paper
Scissors

1. Cut a strip of paper about 3 cm x 16 cm. Hold one end against your chin just below your bottom lip. Blow straight ahead. Does the paper rise up to be almost horizontal?

This simple experiment shows how to create an upward pressure, or lift. According to the Bernoulli principle, the moving air from your breath has less pressure than the still air below the paper. Therefore, the air below pushes up into the lower pressure area and lifts up the paper. In the same way, air passing over an aeroplane wing creates lift. The faster the aeroplane moves through the air, the greater the force of lift.

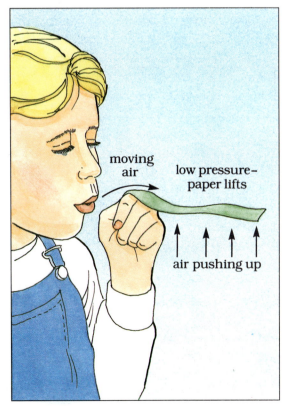

moving air

low pressure– paper lifts

air pushing up

Wings and control

Aircraft wings are specially designed shapes called aerofoils. The top of the wing is curved while the bottom is flat. Aerofoils are shaped like this:

air moving faster

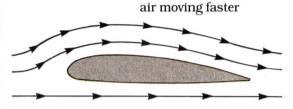

As the wing, or aerofoil, moves through the air, the air separates at the front edge of the wing. Some air passes over the top surface of the wing, and the rest passes along the lower surface. The air moving over the top curved surface has further to go, and, therefore, it travels faster. From the Bernouilli principle we know that the greater the speed of the air, the lower its pressure. This means that the air on top of the wing is at a lower pressure than the air below the wing. This difference in pressure causes the wing to lift.

The angle at which the aircraft's wing meets the oncoming air is also important. The pilot has to ensure that this 'angle of attack' gives the best possible lift. If the angle of attack becomes too steep, the air on top of the wing begins to swirl. This swirling is called turbulence, and it destroys lift, causing the aircraft to stall. When an aeroplane starts to stall, the pilot must act quickly to correct the angle of attack.

An aircraft has three main control surfaces that can be moved to change the direction of flight. These are the rudder, ailerons and elevators.

The rudder is a movable part that is fitted to the vertical tailfin. This

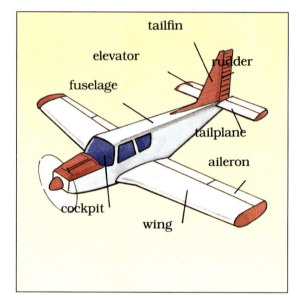

rudder flap is used to steer the aircraft in the air.

The ailerons are two movable flaps on the wings. They can be moved up and down so that the aircraft can be tilted from side to side when it is flying. This movement is called banking and helps the plane change direction in the air.

The elevators are flaps fitted to the tailplanes. These are used at take-off and landing to lift and lower the nose of the plane. When the elevators are raised, the plane climbs. When they are lowered, the plane dives.

Make an aerofoil

You need:
Paper
Sticky Tape
Scissors
Glue
60-cm thread

1. Make a simple working model of an aeroplane wing. Cut a piece of paper 22 cm by 16 cm. Measure 12 cm down each edge and make a dotted line. Make a fold on the 12-cm line. Use the sticky tape to fasten the edges together as shown. The top of the wing should be curved and the bottom flat.

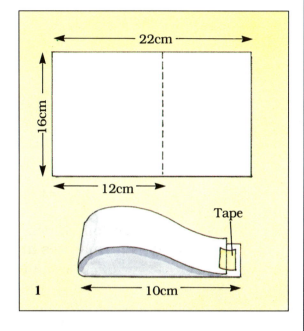

2. Make a small paper rudder and fix it in the centre of the trailing edge.

3. Fix a drinking straw through the thickest part of the aerofoil. Glue it in place. Cut 60 cm of thread and feed it through the straw. Hold each end of the thread and swing your aerofoil through the air. The aerofoil should work its way up the thread. Can you explain why this happens?

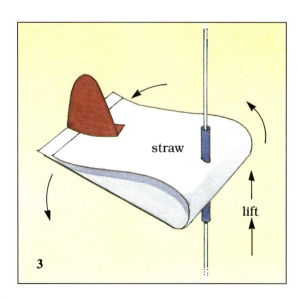

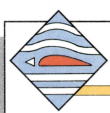

Birds and flight

This albatross, with its long wingspan, glides easily through the air.

The very first birds could not fly well. Their bones were the wrong shape and their wings were not very strong. Over millions of years the body structure of birds has become better adapted for flight. A bird's skeleton is extremely strong but light in weight. The bones are mostly hollow, air-filled tubes. A stream-lined body shape helps many birds to fly long distances.

A bird's wing is made of strong feathers called flight feathers. These are attached to long wing bones, which are moved during flight by very powerful breast muscles. By moving these feathers in various ways, a bird is able to control its flight. The feathers are closed as the bird pushes its wings downwards. The air flows rapidly over the wings causing an upward thrust, or lift. For the upstroke, the wing are lifted with the feathers slightly opened to let the air through. This flapping action pushes the bird through the air. The tail feathers help the bird to turn and change its speed.

Birds with long wingspans, such as the albatross, can take advantage of wind currents for long distances without a wing beat. Hummingbirds can hover in one place by rapidly moving their small wings backwards and forwards in a circular movement.

Watch birds flying

1. Watch some birds flying. Look carefully to see how they flap their wings. Do they glide or hover? Do they fly in straight lines, or do they rise and fall as they move through the air?

2. Watch some birds landing. Do they 'brake' with their wings? How do they use their tails? Do they stand still after landing or do they run along the ground?

3. How do birds take off? Watch how they take off from the ground, from branches, and from the water. Make sketches of your observations.

4. Collect some bird's feathers and try to figure out which part of their bodies they come from. Use a hand lens or microscope to look at the feathers. Make some sketches.

Flying animals

The first flying animals were insects. They developed their wings millions of years before bats or birds. From fossils, we know that insects were flying over 300 million years ago.

Insect wings are very thin, like cellophane. They have many ribs, called veins, that help to stiffen the wings. Flying insects move their wings by using muscles in the body, which pull up and down on the inside of the wings. Most insects have two pairs of wings, but some insects, such as the housefly, the hoverfly and the mosquito, have only one pair. The hind wings have developed into stumps that help the insect balance itself when in flight.

Hoverflies are expert fliers. By moving their wings extremely fast, they can fly forwards, hover in one spot, and even fly backwards. Insects fly at different speeds. The housefly's average speed is about 8 kph, while the dragonfly, which is one of the fastest flying insects, can fly almost 50 kph.

Bats are the only mammals that can truly fly. Their large wings are formed by elastic skin stretched over their long finger bones.

There are flying fish that can glide above the surface of the sea for a few seconds. They take off from just below the surface by using the tail fin. The large pectoral fins spread out like wings as the fish move through the air.

The dragonfly is one of the fastest flying insects.

Make a model insect

You need:
Tissue paper
Plastic bags
Cardboard
Glue
Scissors
Straws
Thin wire
Paints

1. Look carefully at some photographs of flying insects. Use materials such as those listed above and others that you have at home to make an accurate model of an insect with wings.
How will you make the body?
What shape will the wings be?
How many pairs of wings will your insect have?
Will the wings be in an upright or flat position?
What materials will you use to make the wings?

2. Finish your model by carefully painting it.

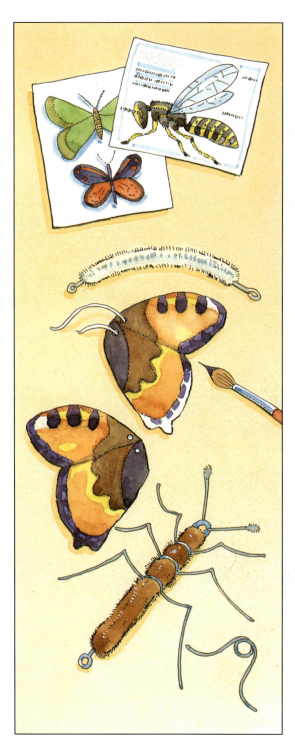

Parachutes and gravity

Objects fall to the ground because they are pulled towards the earth by a force called gravity. Find out more about gravity by doing these tests:

1. Drop two small pebbles from shoulder height at the same moment. Do they hit the ground at the same time?

2. Repeat this test with one pebble and a tightly crumpled-up piece of paper. The pebble and the ball of paper will hit the ground at the same time because the pull of gravity is the same on all objects, no matter how heavy they are.

3. Try the same test with a flat piece of paper and a crumpled-up piece of paper. Repeat this test a few times. Why do you think the flat paper falls more slowly?

A parachute making a slow descent.

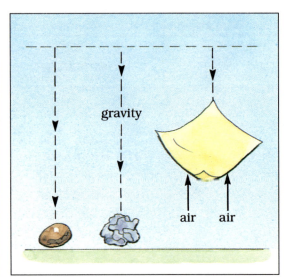

As the papers fall, the air passing around them acts as a break and slows down their fall. The flat paper falls more slowly than the crumpled-up one because its larger area means that more air is trapped underneath. This breaking force of the air is sometimes called drag. Parachutes use this breaking force to bring flyers safely down to earth. Some fast jets use parachutes to slow down as they land.

Making a parachute

You need:
4 lengths of string (40 cm each)
Plastic bags
4 metal washers
Scissors

1. Cut a 25 cm square of plastic. Tie the four lengths of string to the corners of the plastic square. Attach a metal washer to the ends of the string as shown.

2. Fold your parachute into a ball and throw it high into the air. Does it open and float gently to the ground?

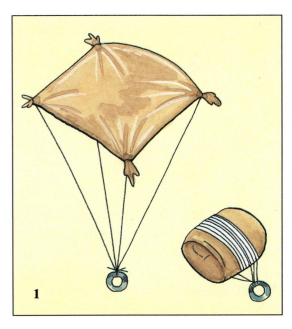

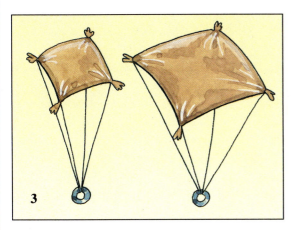

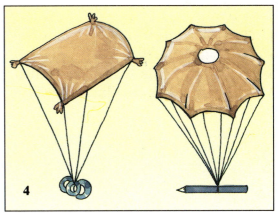

3. Try making several different-sized parachutes. Use the same material and keep the string and the washers the same size. Does the size of the parachute change the way it falls to the ground?

4. Design and make some parachutes using different materials and varying the number of washers. A modern parachute has a hole at the top to let some of the air out and stop it from swaying from side to side. Will this work with your parachutes?

Seeds in flight

The fine hairs on this dandelion seed act like a parachute.

If all the seeds made by a plant were to fall on the same bit of land, the new plants would be too close together to grow properly. Plants use a variety of methods to disperse their seeds. The dandelion has fine hairs fixed to each seed. These hairs act as a parachute to carry the seed many kilometres away from the parent plant.

Collect some dandelion seeds and look at them carefully, using a hand lens or microscope. How are the hairs arranged to form the parachute? Draw a parachute. Are all the dandelion parachutes the same size? Let some parachutes fall. Do they all reach the ground at the same time?

The winged fruits of the sycamore also travel through the air. Collect some sycamore fruits. Look at them closely. Throw some into the air and watch how they fall to the ground. How far will they travel on a windy day?

Make paper spinners

You need:
Several pieces of paper
Ruler and pencil
Paperclips
Scissors

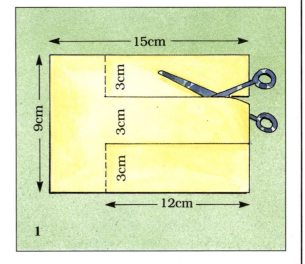

1. Cut a piece of paper to measure 15 cm by 9 cm.
   Make two cuts, each 12 cm long, as shown.

2. Fold the end pieces as shown and attach a paperclip to the centre piece.

3. Drop your spinner and watch what happens. Try adding more paperclips. What happens if you shorten one of the rotors? What happens if you drop your spinner upside down?

Design and make
Make some large spinners using other materials.

Lighter than air

In 1783, two French brothers, named Montgolfier, built a very large balloon from paper and linen. They heated the air inside the balloon by building a fire under it. Hot air is lighter than cold air, so as the air inside the balloon was heated, the balloon began to rise up into the air. Later that year the Montgolfier brothers launched a hot-air balloon with two passengers inside a basket carrier attached to it. This was the first successful human flight.

Ballooning is now a popular sport. The balloon is often made from nylon or polyester with strong webbing reinforcements. The balloon pilot switches on a burner to heat the air inside the balloon and make it rise. When the burner is switched off, the air cools and the balloon gradually descends. The pilot has little control over the direction of the balloon's flight; the balloon is simply carried along with the wind.

Airships are gas-filled balloons fitted with engines and propellers to help control the direction of their flight. Early airships were filled with hydrogen, which is lighter than air, and used for passenger flight. However, hydrogen is very flammable. In 1937, the German airship the *Hindenberg* hit a tall mooring mast as it was landing in New Jersey, USA, after its transatlantic flight. The airship burst into flames, killing thirty-five

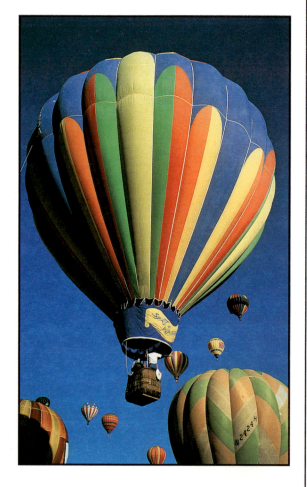

The air inside a hot-air balloon is heated with a burner.

of the ninety-seven people on board. This tragedy marked the end of the airship as a form of passenger transport. The future lay with the aeroplane.

Today's airships are filled with helium, which is less flammable than hydrogen. They are used mainly for advertising, aerial surveys and pollution monitoring.

Make a hot-air balloon

You need:
A large plastic bag
Some thin string
A small plastic container such as a
   yoghurt pot
A hairdryer (adult supervision needed)

Note: Make sure that your plastic bag is made of very thin plastic and has a large volume.

1. Tie the container to the plastic bag with string as shown.

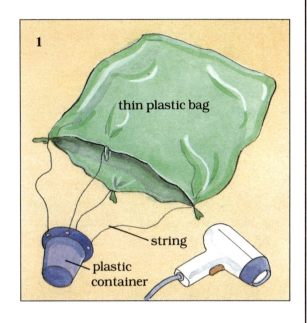

thin plastic bag

string

plastic
container

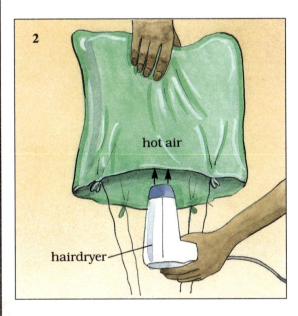

hot air

hairdryer

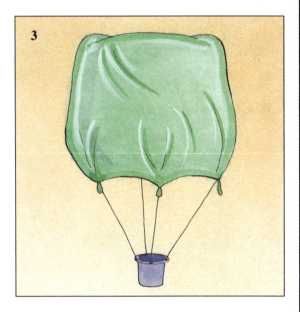

2. Ask a friend to hold the plastic bag open while you inflate it with hot air using a hairdryer. Hold the blower just below the opening and not too close to the plastic.

3. Release the balloon. How long does it stay in the air? Did it rise quickly or slowly?

Kites

People have been flying kites for about four thousand years. The first kites were made in China and Japan and were probably developed from banners that streamed in the wind.

Kites have been used to lift people into the air to spy on enemies during wartime and to carry observers aloft at sea. They have also been used for photography and to help with weather forecasting. Today, in a sport called parascending, a person is attached to a large kite that is towed behind a speedboat. The lift produced by the kite causes it to rise high into the sky.

Many people enjoy flying kites. The kite is controlled by a line held by someone on the ground. Some kites can be made to swoop and soar up and down through the air. The line holds the kite at an angle to the wind, so the air pushes upwards against its under-surface. Some kites act like a wing; as the air moves over the kite, lift is created.

This kite is about to take to the air.

Make and fly a kite

You need:

A light plastic bag

Two thin bamboo sticks or dowel rods

A reel of string

Sticky tape or staples

Warning: Never fly kites near overhead wires.

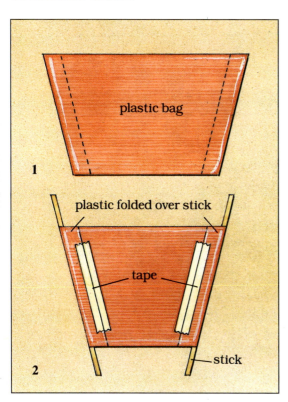

plastic bag

plastic folded over stick

tape

stick

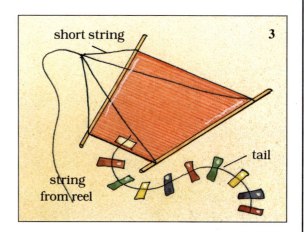

short string

tail

string from reel

3. Measure four short pieces of string and tie them as shown. Tie your reel of string to the four short strips. Make a long tail from strips of plastic. Tape this tail to the bottom of the kite.

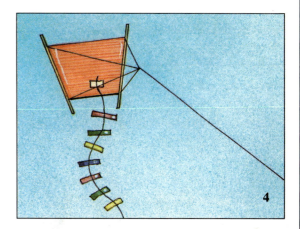

1. Cut the plastic bag into the shape shown in the diagram.

2. Position the two sticks near the sides of the plastic. Fold the plastic over the sticks and tape or staple the edges.

4. Launch your kite in a steady wind. Does it help if you walk into the wind and jerk the string. Why does your kite need a tail? Why does your kite rise into the air?

Fly your kite in a large open space; kites are easily entangled in trees.

Gliders

This high-performance two-seater glider soars above the countryside.

In the nineteenth century, an Englishman, Sir George Cayley, studied the flight of birds and noticed that some birds, such as seagulls, could glide in the air for long distances without moving their wings. Cayley believed it was possible to build a machine that could glide through the air in a similar way. In 1853, he succeeded in building a glider with three pairs of wings that carried his coachman a short distance in the air. This Cayley glider was the first heavier-than-air aircraft to carry a person aloft.

Balloons and airships can leave the ground because they are filled with a gas that is lighter than air. A glider, however, needs help to overcome gravity. A modern glider is often towed along behind an aircraft to get it into the air. The forward movement causes air to move faster above the glider's wings than below. This generates lift (see chapter 2). The glider pilot takes advantage of rising air currents to stay in the air. The rising air currents push the glider higher and make its flight last longer.

A glider is built of light materials, and its long, thin wings and narrow fuselage are designed to increase lift and reduce drag.

Gliders have been used in war to carry troops into action. Today, gliders are used to help in training pilots. Gliding has also become a very popular sport.

Build a glider

You need:
2-mm-thick balsa wood or very thin
   polystyrene
Modelling knife
Glue
Blu Tac or paper clip

1. Copy the outlines of the plane parts on to the balsa wood. Check that they lie in the correct direction with the grain of the wood.

   Carefully cut out the shapes with a sharp modelling knife.

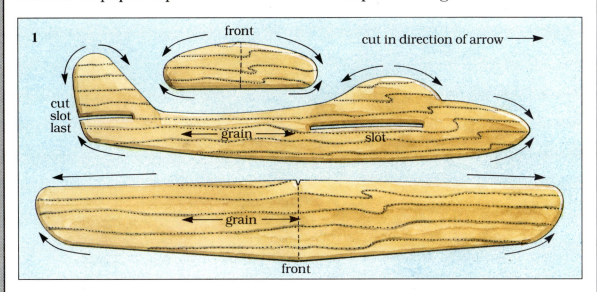

   Do the cutting on top of a cutting board. Where possible cut across the grain rather than along it; this should help prevent splitting. See guide arrows on diagrams.

   Slide the pieces together and fix with tiny dots of glue on each part. The model glider must now be balanced. Do this with either a paper clip or some Blu Tac. See diagrams.

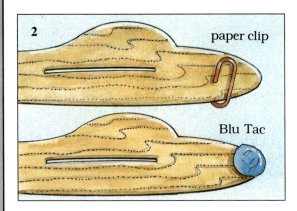

2. Test fly the model by holding under the main wing and throwing forward. If the model stalls, add a little more Blu Tac or move the paper clip forward. If the model nose dives, remove a little Blu Tac or move the paper clip backwards.

   Test fly again. Make further adjustments as needed until a good glide is obtained.

Hang gliders

A hang-glider pilot is suspended underneath a large wing.

Hang gliding was started in 1891 by a daring German inventor named Otto Lilienthal. He constructed a large set of wings connected to a tail, made of willow rods covered with fabric. The pilot hung from a frame in the centre of the wings. Lilienthal took off in his glider by running downhill until he was moving fast enough for the wind to lift the craft. Once airborne, he controlled his glider by shifting his weight from side to side. On his first flight, he was able to fly 250 m using this method. Lilienthal built large gliders and made over 2,000 flights in them. He was killed in 1896 while testing one of his new gliders.

A modern hang glider pilot is strapped into a frame that hangs from a large nylon wing. The pilot lies horizontally holding a control bar. The pilot launches the glider in the same way that Lilienthal launched the first glider – by running down the side of a hill. Once in the air, the pilot steers the glider by pushing his or her weight against the control bar.

Make a hang glider

You need:
Four drinking straws
A plastic bag
Some sticky tape
Thread
Plasticine
Scissors

1. Arrange the straws in a triangle shape. Stick the corners together using the tape.

Cut and position another straw in the centre as shown. Again use the sticky tape to fix it in place.

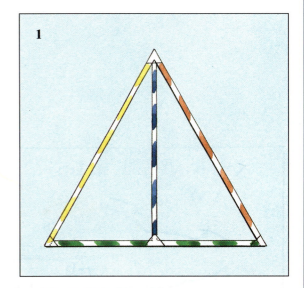

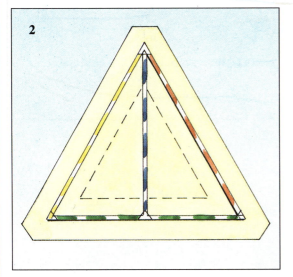

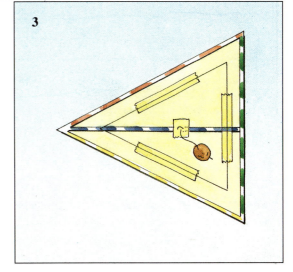

2. Cut a piece of plastic to fit over the straw framework. Stick the overlap down over the straws.

3. Loop a short piece of thread over the central straw as shown. Attach a small piece of plasticine to this cotton loop. Fix the loop in place with sticky tape. Test your hang glider. How far does it fly? You may need to alter the position and size of your plasticine weight.

Propellers

Aeroplane propellers are designed to push the air backwards, thereby forcing the aeroplane forward through the air. The first aeroplanes used propellers, and propeller-driven aircraft are still in use today.

The propellers used on aeroplanes are similar to those used on ships, but aeroplane propellers need to revolve at a greater speed. Many aeroplane propellers have three blades. Each blade has an aerofoil shape similar to that used on aircraft wings. As the blades revolve, their shape produces a lifting effect and thrusts the aircraft through the air. In the early days, propeller blades were made from solid wood. Modern propellers are made from metal

Some modern planes are propeller-driven. As the propeller blades revolve in the air, their aerofoil shape creates a lifting effect.

alloys or plastic materials.

The flatter the angle of the blade, the smaller the bite into the air. Smaller bites give high pulling power but low speed. The blades of propellers can be partly rotated so that they have flat angles on take-off and steep angles for normal flight. In early planes, the angles of the blades were changed by controls operated by the pilot. In modern aircraft, the angles of the blades change automatically.

Rubber band power

You need:

A plastic propeller (from a model shop)
A plastic pipe (about 20 cm long)
A small bead
Some elastic bands
Nylon fishing line

A cork
2 paperclips
Strong wire
Scissors
Pliers
A small piece of wood

1. Make a hole through the centre of the cork. Push a paperclip through this hole and then through the bead and propeller as shown. Bend the end of the paperclip over the propeller hub.

2. Tie the elastic bands together and fix them to the paperclip. Slot the elastic bands through the plastic pipe and fit the cork in the end of the pipe. Use a small piece of wood to hold the elastic in place.

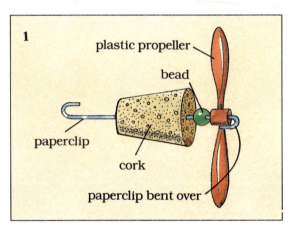

1
plastic propeller
bead
paperclip
cork
paperclip bent over

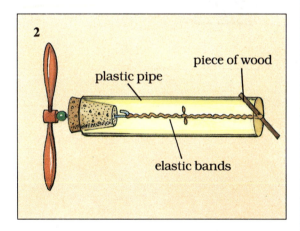

2
piece of wood
plastic pipe
elastic bands

3. Cut two equal lengths of wire. Each one must be long enough to wrap round the pipe and extend out beyond the propeller. Twist the wires around the plastic pipe and make a hook at the free end of each one.

Stretch the nylon line across a room and fix it to two supports. It must be taut. Hang your machine on this line. Turn the propeller about 20 times and let it go.

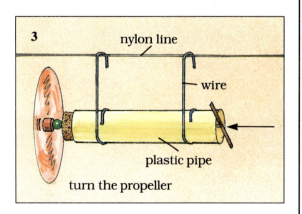

3
nylon line
wire
plastic pipe
turn the propeller

Aircraft engines

Powered aircraft need engines in order to take off and stay in the air. Some early flying pioneers tried using steam engines, but without much success. Aircraft builders needed engines that were light and reliable, but also powerful enough to get aircraft into the air.

The solution to this problem came with the invention of the petrol engine, originally developed for use in the motor car. The aircraft *Wright Flyer I* made the first controlled powered flight as Kitty Hawk in the USA on 17 December 1903. It was powered by a small petrol engine designed and built by Orville and Wilbur Wright.

The Wright brothers' engine had four cylinders arranged in a straight line. The pistons were linked to a crankshaft on the end of which the propeller was fixed. Soon, other engines designs began to appear. In radial and rotary engines the cylinders are arranged in a ring around the crankshaft.

Like all internal combustion engines, aircraft engines get very hot and must be cooled. Some of the early engines were water-cooled, but others were equipped with large fins and were cooled by air. An air-cooled engine is light, but it does need a large surface area exposed to the rushing air. But, this has the disadvantage of increasing drag and slowing the aircraft down.

The Wright brothers designed and built a petrol engine for the Wright Flyer I.

Make a powered aircraft

You need:

Engine unit (see page 27)
Balsa wood
  2 pieces 600 x 75 x 3 mm
  2 pieces 400 x 6 x 10 mm
Rubber band
Balsa cement
A wing fixing (from a model shop)

1. Make wings, tail planes and tailfin of the glider from balsa wood 3 mm thick (see diagram). Use the 10-mm strips for fuselage strips.

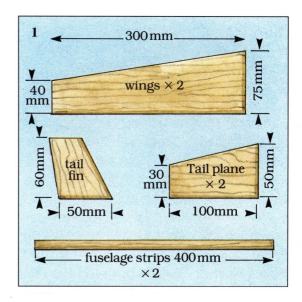

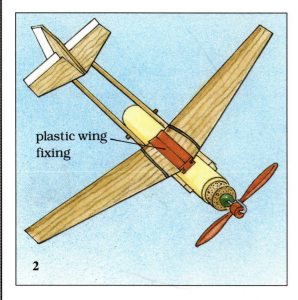

2. Assemble as shown. Mount the engine unit with a rubber band over the wings. Move the engine forwards or backwards or add a little weight front or rear (Blu Tac or Plasticine) until the plane glides without the engine running.

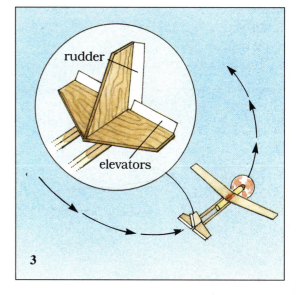

3. When the plane is trimmed, wind the propeller and launch on a level flight path. Experiment with thin paper elevators on the tail plane. Set a rudder at an angle on the vertical fin and the plane will fly in a curved path.

Helicopters

A helicopter is an aircraft that can take off vertically, hover in mid-air and fly in any direction. The helicopter gets its lift from horizontal rotors fixed to a revolving shaft. A powerful engine turns the shaft. The rotor blades of a helicopter use the same system as the wings of an aeroplane to obtain lift. When the rotor blade assembly is tilted slightly, the helicopter is able to move forward. When the rotors are spinning, they tend to make the helicopter's fuselage turn in the opposite direction. To counter this, many helicopters have a small tail rotor. This prevents the fuselage from turning.

Helicopters are particularly useful because they can take off and land in a small space and do not need a runway. Special landing and take-off areas, known as heliports or helipads, can be found on oil-drilling rigs and on top of skyscrapers. Helicopters are often used to rescue people from the sea and to carry soldiers into battle. After space flights, astronauts have been picked up from the oceans by helicopters and flown to nearby ships. Wherever there are disasters, such as earthquakes and floods, helicopters are in action, rescuing people and flying in food and medicine.

Helicopters are often used for rescue missions.

Make a model helicopter

You need:
A length of dowel
A plastic lemonade bottle
Scissors
A nail

1. Cut a rectangular strip of plastic from the bottle. This is going to be the rotor blade.

2. Use a nail to make a hole in the centre of the plastic strip.

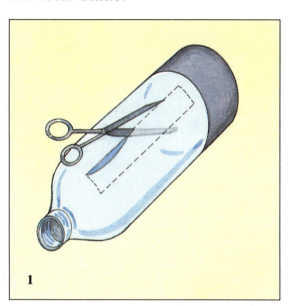

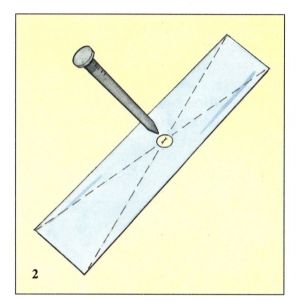

3. Push the dowel through the hole in the plastic strip. It must be a tight fit. Twist the blades slightly. Now you are ready for a test flight. Spin the helicopter with your hands and throw it into the air.

How can you improve your helicopter design? Try making rotor blades that are short and fat. What happens when the blade is long and thin?

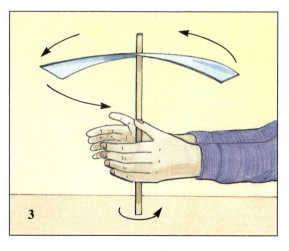

Jet power

The invention of the jet engine, by the British engineer Frank Whittle, in 1937, changed the pattern of air travel. Aeroplanes fitted with jet engines are able to fly extremely fast and at great heights.

The turbojet was the first form of jet engine. Air is drawn in at the front and compressed (squeezed) by a turbine. This compressed air then passes into a combustion chamber where it heats up. Fuel is injected into the hot air. It burns, and the hot gases that are produced expand and shoot out through a nozzle at the back of the engine, thereby thrusting the aeroplane forward. These gases also turn a turbine wheel that drives the air compressor. Turbojets are used in modern fighters and other high-speed aircraft.

Another type of jet engine is the turboprop. Here the hot gases work a turbine that turns a propeller.

Most of today's airliners are powered by turbofan engines. A turbofan has a large low-pressure fan in front of the compressor turbine. Part of the airflow generated by this fan flows through the engine, the rest flows round the outside. This type of jet engine is more efficient at low speeds and is quieter than a turbojet.

Jet engines were first fitted to military aircraft towards the end of the Second World War. They were able to fly much faster than

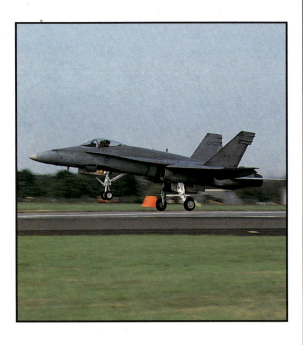

Jet engines provide enormous power.

propeller-powered fighter aircraft. Later, jet engines were used in passenger aircraft, and today, aircraft such as the jumbo jets carry thousands of people daily all over the globe. Some military aircraft, such as the British Harriers, can adjust their jet engines for vertical take-off and landing.

The introduction of jet engines and high-speed travel has meant that aircraft designs have changed. Stronger materials have been introduced for the construction of aircraft. Streamlined shapes have been developed and many changes have been made to the controls that pilots use to handle these large aircraft.

A jet-propelled cable car

You need:

A balloon
Nylon fishing line
Sticky tape

A drinking straw
Some paper
Scissors

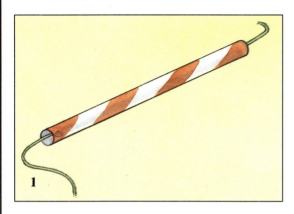

1. Pass the nylon line through the straw. Fasten the nylon line to two supports so that it is taut.

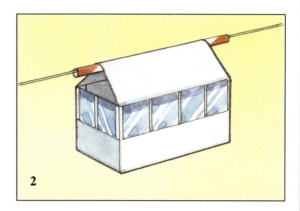

2. Design and make a simple cable car from paper and fix this to the straw on the line.

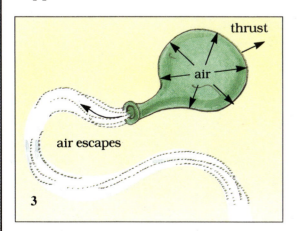

thrust

air

air escapes

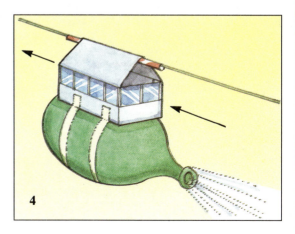

3. Blow up a balloon and hold the neck. Watch how the balloon flies through the air when you let it go. As the air rushes out, the balloon is thrust forward in the opposite direction.

4. Use the balloon to make the cable car move along the line. How far will your cable car travel?

Supersonic flight

When an aeroplane flies, it pushes air out of its way. Normally, the displaced air flows smoothly over the wings. But when an aeroplane approaches the speed of sound (1,065 kph at 10,500 m above sea level, sometimes called Mach 1), the air particles do not have enough time to move out of the way. They pile up along the front edge of the wings, causing the air around the plane to become extremely turbulent. The aeroplane is said to come up against a 'sound barrier'. Pilots were sometimes killed when their aircraft became uncontrollable as they approached the sound barrier.

In 1947, 'Chuck' Yeager of the USA became the first person to fly faster than the speed of sound. His aircraft was specially built to travel at supersonic speeds and was powered by a rocket engine. When Yeager accelerated to Mach 1, he experienced terrific shaking, but he found that past the sound barrier the flight was smooth. When Yeager's plane travelled faster than the speed of sound, it moved faster than the disturbances it had made in the air, thus it was able to 'break' the sound barrier.

Aeroplanes are specially designed to break the sound barrier without being shaken to pieces by turbulence. Supersonic aircraft have pointed noses and swept-back or triangular delta wings to reduce drag.

The Anglo-French Concorde can fly twice as fast as the speed of sound.

Make a delta-wing glider

You need:
Large polystyrene food tray
Glue
Sticky tape
Scissors
Thin card
Plasticine

1. First cut out the delta wing shape, then cut out the fuselage shape. Glue the wings and fuselage together. Test your glider. Does it travel in a straight line? Add plasticine weights to the nose until the glider travels along a straight flight path.

Polystyrene is light but quite strong and, therefore, a good material to use for model gliders.

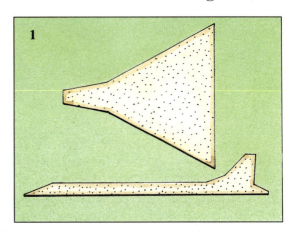

2. Cut six small pieces of card and use sticky tape to attach them to the wings as shown. Add a piece of card to act as a rudder.

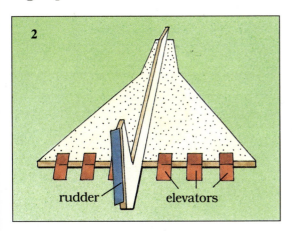

rudder   elevators

3. Set the elevons (combined ailerons and elevators) to an up position and launch your glider. What happens? Now set the elevons to a down position and launch your glider. What happens? What happens when the elevons on one side are raised but lowered on the other?

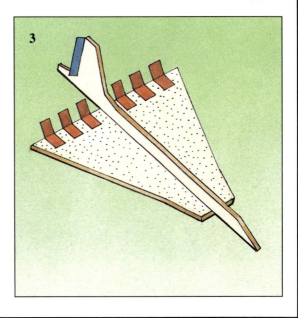

Passenger flight

Aircraft have two main uses. They can be used as war machines or as a form of transport. Today, the speed with which aircraft can travel round the world has made them an important means of transporting passengers.

Passenger transport began soon after the First World War. By then people had realized how useful aeroplanes could be, and aircraft pioneers were starting to show that long-distance flights could be made safely. In 1919 civil airlines opened up in Britain, France and Germany. By 1929 an England to India service was in operation, and in 1935 an England to Australia service was opened.

The first passenger aircraft were modified wartime bombers, such as the de Havilland D.H.4a and the Vickers Vimy. However, large purpose-built passenger aircraft, such as the Farman F.60 Goliath, soon began to appear. Seaplanes were commonly used, since in many places there were no suitable landing strips on land. Huge flying boats became a popular means of long-distance transport throughout the world.

Most of the early passenger aircraft were biplanes (two sets of wings) or monoplanes (one set of wings) with the wings set high on the fuselage. In the mid-1930s, however, a new design of aeroplane began to

This light aircraft is used by the Flying Doctor Service in Australia.

appear. Aircraft such as the Boeing 247 and the Douglas DC-2 were all-metal, low-wing monoplanes – a design that is still in use today.

During the Second World War there were several developments that helped to improve passenger flight. Aircraft became larger, and engines became more powerful and more reliable. Radar became available, and radio navigation aids had been greatly improved. Soon after the war, the first jet airliners were introduced. The first turbojet airliner, the de Havilland Comet, appeared in 1952. It had a cruising speed of 788 kph, which was over 250 kph faster than the best turboprop airliners of the time.

Turbofan engines were also introduced during the 1950s, and

Aircraft provide daily passenger services around the world.

airliners now had pressurized cabins for flying at high altitudes. More and more people wanted to travel by air, and the first wide-bodied aircraft, such as the Boeing 747 and the McDonnell Douglas DC-10 appeared in 1969. The Boeing 747 can carry up to 500 passengers, and the latest version, the 747-400, is a fully computerized aircraft with digital displays rather than dials in the cockpit. Smaller airliners, such as the Boeing 767 and the Airbus A320, are also equipped with electronic fly-by-wire systems (see page 39). Concorde, introduced in 1969, remains the only supersonic airliner in service.

Flight control

Early pilots had few aids to help them find their way. A compass could be used to find direction, but pilots often had to navigate using landmarks such as towns and railways. Modern pilots, on the other hand, have a variety of sophisticated instruments to help them fly their aircraft.

Unlike the drivers of land and sea vehicles, an aircraft pilots needs to know how high the aircraft is off the ground. An altimeter provides a pilot with this information and was one of the first instruments to be introduced. An airspeed indicator tells the pilot how fast the aircraft is moving through the air. An instrument called an artificial horizon keeps the pilot informed of the aircraft's attitude (the degree to which the aircraft is tilted to the right or left). A gyrocompass works along with a magnetic compass to make sure that the aircraft continues to head in the right direction, and another set of instruments keeps a check on the rate at which the aircraft is being blown sideways by the wind.

An artificial horizon and a gyrocompass both contain gyroscopes. A gyroscope is a useful device because the spinning wheel at its centre tends to resist being moved from side to side; that is, it tries to remain pointing in the same direction all the time. When linked

The control tower of a busy airport. Radio signals are used to help the aircraft land in poor visibility.

to appropriate instruments, gyroscopes can therefore be used to sense aircraft movements. Three gyroscopes, each one spinning in a plane at right angles to the other two, form the basis of an autopilot that can be set to keep an aeroplane flying straight and level. A modern autopilot is controlled by a computer, which can be programmed to keep the aircraft accurately on course.

Gyroscopes also form part of an aircraft's inertial navigation system. This charts every movement the aircraft makes and calculates the aircraft's position in relation to a

Modern aircraft have sophisticated flight control systems.

known starting point. Other navigation systems use radio signals beamed from ground stations. Automatic direction finding (ADF) systems collect signals from three stations in order to find the aircraft's position. Alternatively, radio signals from satellites can be used to pinpoint an aircraft's position very accurately. Airliners often follow radio tracks produced by ground-based stations. And at airports, radio beacons are used to control air traffic and help aircraft land in poor visibility.

Until quite recently, the control surfaces (such as ailerons and elevators) of an aircraft were operated by cables or hydraulic fluids linked directly to the pilot's control stick. In some modern aircraft, these links have been replaced by much lighter electrical wires. The control surfaces are operated by small motors, which receive digitally coded electrical signals from the aircraft's computer. This in turn receives signals from the pilot's controls. The system is known as 'fly-by-wire' and is designed to make aircraft lighter and easier to fly.

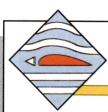

Space flight

Thanks to the invention of the rocket engine, people are now able to escape the pull of the earth's gravity and travel into space. Rockets are much faster than the most powerful jet, and they can function in airless space because they carry their own oxygen supply. Fuel burns in the combustion chamber, creating very hot gases. The gases expand and thrust the rocket forwards.

The Space Age began in 1957 when the Russians put a satellite into orbit around the earth. Since then many other satellites have been placed in orbit to help with weather forecasting and communications. In 1969 American astronauts were blasted off from Cape Kennedy, USA, and made the first manned landing

The launching of the US Space Shuttle is a magnificent sight.

on the moon.

The US Space Shuttle has been designed to be used over and over again. It consists of four main parts – an orbiter shaped like an aeroplane, two rocket boosters and a large fuel tank. At lift-off, the rocket boosters help lift the orbiter into the clouds, but they, and the fuel tank, are jettisoned as the orbiter heads into space. After completing its task in space, the orbiter fires retro-rockets and re-enters the earth's atmosphere. It glides to a touchdown on an ordinary runway. Within weeks it can be ready for another flight.

Make a water rocket

You need:

2 empty, clean plastic drink bottles
Rubber bung or cork to fit neck of
  bottle
2 elastic bands
Football inflator valve
Several bicycle pump connectors

String
Wooden stake
Hammer

Vaseline
Scissors
Bicycle pump

1. To make a rocket shape, cut the top from one of the plastic bottles. Fit this section on to the base of the second bottle, as shown, to form a nose cone. Cut four fins from the remaining section of the first bottle. Use two elastic bands to fasten the fins around the base of the rocket body.

2. Make a hole through the centre of the rubber bung. Fit the inflator valve through the hole and screw the valve to a number of bicycle pump connectors, to form a long tube. Connect the end of this tube to the bicycle pump.

3. Hammer the wooden stake into the ground. Use string to attach the bung and connectors to the top of the stake. Smear the bung with Vaseline. Pour a cupful of water into the rocket body. Carefully press the rocket base on to the rubber bung. Try to spill very little water.

4. Pump air into the rocket until the water is forced downwards and the rocket takes off at high speed.

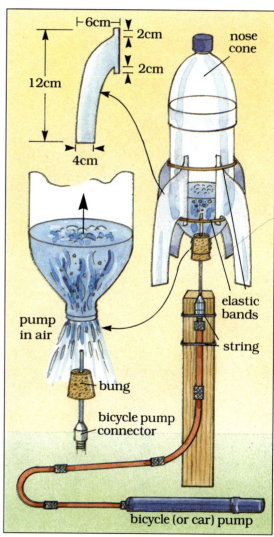

6cm
2cm
2cm
12cm
4cm
nose cone
pump in air
bung
bicycle pump connector
elastic bands
string
bicycle (or car) pump

WARNING: This activity must be supervised by an adult. Do not point the rocket at people or buildings. Stand well back and work outside.

The next steps

Since the first successful powered flight of the Wright brothers in 1903 the aeroplane has undergone many changes. And the technologies involved in building and flying aircraft continue to change rapidly.

For example, a variety of new light-weight materials are being used. New materials include titanium, special kinds of steel and an alloy of aluminium and lithium, which is the lightest known metal. Composites – materials made up of mixtures of two or more different substances – are being developed for a wide range of uses. Carbon fibre composites (fibres of carbon embedded in plastic) have been in use for a number of years. But today, scientists are developing silicon carbide fibres embedded in titanium, as well as mixtures of aluminium and titanium.

Ceramics are also playing an increasing part in aircraft construction, and in the future these highly heat-resistant materials will probably be used in aircraft engines. Future engines will be more reliable and will burn fuel more efficiently.

Computers will, of course, continue to play an important part in the design, construction and flight control of aircraft. Fly-by-wire systems could soon be replaced by

The B-2 Stealth Bomber is almost invisible on radar screens.

A design for a supersonic aircraft of the future.

fly-by-light systems, in which signals are transmitted around the aircraft along optical fibres (thin strands of glass) instead of electrical cables. An optical fibre can carry a great deal more information than an electrical wire, so fly-by-light systems can be made considerably lighter. Aircraft manufacturers are also investigating systems in which some aircraft functions can be controlled by pilot voice commands.

As the number of people wishing to travel by air continues to increase, future aircraft may be even larger than today's wide-bodied aircraft. The next generation of jumbo jets may carry up to 700 passengers.

Some military aircraft of the future may be little more than flying wings. These so-called Stealth aircraft have been designed to be almost invisible to enemy radar. Constructed of composite materials, they have no vertical surfaces or sharp edges, and they are painted with a special paint designed to confuse radar signals.

Space planes are also being considered for the future. They are currently two designs for such aircraft, the British Hotol and the American Orient Express. This type of aircraft will take off normally, but will then be boosted into space by powerful engines, reaching a speed of about 20 times the speed of sound. As it nears its destination, it will descend into the atmosphere. Such aircraft should be able to fly halfway round the world in less than an hour.

Glossary

Acceleration Rate of increase in speed.

Aerofoil A shaped surface, such as a wing on an aeroplane, that is designed to give lift when air flows past it.

Aeroplane A heavier-than-air aircraft with fixed wings.

Aircraft A machine that travels through the air. Airships, balloons, helicopters and aeroplanes are all types of aircraft.

Alloy A material made from a mixture of two or more metals.

Crankshaft A shaft within an engine that has cranks that change the to-and-fro movement of a piston into the turning movement of the mainshaft. This then drives the propeller of an aeroplane.

Cylinders Tube-shaped spaces inside an engine block. Pistons move to and fro inside the cylinders.

Drag The slowing down effect that air has on an object moving through it.

Elevon A combined aileron and elevator, used on a delta-winged aeroplane.

Fuselage The body of an aeroplane.

Glider An aircraft with wings but no engine.

Gravity The force that tends to draw all bodies on the earth's surface or in the earth's sphere towards the centre of the earth.

Helium A very light, colourless, odourless gas. Helium is the second lightest gas.

Hydrogen A very light, colourless, odourless gas. Hydrogen, the lightest gas, is extremely flammable.

Internal combustion engine An engine in which fuel burns inside a cylinder and so drives a piston.

Lift The upward pressure that air exerts on an aerofoil or wing.

Lift-off The moment when a rocket rises from the launch pad.

Orbit The path of a planet, spacecraft or artificial satellite around a planet or star.

Satellite An artificial satellite is a man-made object that circles a planet in space. The moon is the earth's natural satellite.

Stall An aircraft stalls when it loses lift and begins to fall out of control.

Thrust The force produced by an engine to propel an aeroplane or rocket through the air.

Turbulence A disturbance in the airflow that can lead to loss of lift.

 # Further information

Books to read

Allen, J.P. and Martin, R., *Space, the Ultimate Challenge* (Marks & Spencer PLC, 1986).

Barnaby, R.S., *How to Make and Fly Paper Aircraft* (Pan Books, 1973).

Bradbrooke, J., *The World's Helicopters* (The Bodley Head, 1972).

Brown, D., *History Eye-Witness, Flyers* (Hamlyn, 1981).

Catherall, E., *Balls and Balloons* (Wayland, 1985).

Couper, H. and Henbest, N., *Spaceprobes and Satellites* (Franklin Watts, 1987).

Gatland, K., *Young Scientist Book of Spaceflight* (Usborne Publishing, 1976).

Gatland, K., *Exploring Space* (Macdonald Educational, 1976).

Hawkes, N., *Space Shuttle* (Collins, 1982).

Kurth, H., *Helicopters* (World's Work Ltd, 1975).

Lambert, M. *Aircraft Technology* (Wayland, 1989).

Marshall, R. and Bradley, J., *Watch it Work! The Plane* (Viking Kestrel, 1985).

Shepherd, W., *How Aeroplanes Fly* (Hart-Davis, 1971).

Organizations to contact

Air Education and Recreation
  Organization
Carwarden House
118 Upper Chobham Road
Camberley
Surrey

British Balloon and Airship Club
42 Bath Road
Longwell Green
Bristol

British Interplanetary Society
27–29 South Lambeth Road
London SW8 1SZ

British National Space Centre
Millbank Tower
Millbank
London SW1P 4QU

Junior Astronomical Society
36 Fairway
Keyworth
Nottingham NG12 5DU

Royal Society for the Protection of
  Birds
The Lodge
Sandy
Bedfordshire SG19 2DL

Places to visit

Air and Space Museum
Smithsonian Institute
Washington D.C.
USA

Fleet Air Arm Museum
Yeovilton
Somerset

Greater Manchester Museum of
  Science and Industry
Liverpool Road
Manchester

Jodrell Bank
Macclesfield
Cheshire SK11 9DL

Museum of Flight
East Fortune
Haddington
East Lothian
Scotland

R.A.F. Aerospace Museum
Cosford
Wolverhampton
West Midlands WV7 3EX

Science Museum
South Kensington
London SW7 2DD

You could also visit local airports
museums and planetariums, as
well as wildfowl reserves and
butterfly farms. Air shows are
good places to see new and
historical aircraft. They may
include aerobatic displays and
parachute jumping. The
Farnborough (England) air show is
one of the largest and best known.

Notes for parents and teachers

This book will be useful to teachers in implementing the National Curriculum at Key Stages 1, 2 and 3. The information and activities relate to the following:

Technology attainment targets 1, 2, 3, and 4
Science attainment targets 1, 2, 6, 7, 9, 10, 13, 14 and 16
Flight can also be developed as a cross-curricular topic that includes National Curriculum English and Mathematics.

There are many activities in this book that will require the help of a teacher or parent. Parents will also find it helpful to consult the section on places to visit during weekends or school holidays.

Picture acknowledgements

The publishers would like to thank the following for supplying pictures: Civil Aviation Authority 38; Oxford Scientific Films 10, 12, 76; Photri 5, 9, 40, 43; Topham 26, 39, 42; Wayland Picture Library 4, 28, 34, 36, 37; ZEFA *cover*, 14, 18, 20, 24, 30, 32.

Index